FACT FINDERS
KNOW THE DIFFERENCE!

PLANT OR FUNGUS?

EZRA E. KNOPP

PowerKiDS press

Published in 2026 by The Rosen Publishing Group, Inc.
2544 Clinton Street, Buffalo, NY 14224

First Edition

Editor: Greg Roza
Book Design: Michael Flynn

Photo Credits: Cover (plants) Elgub/Shutterstock.com; cover, p. 4 (mushrooms) IH82/Shutterstock.com; p. 5 (tree stump with fungus) Tintila Corina/Shutterstock.com; p. 5 (flower) Fabrizio Guarisco/Shutterstock.com; p. 6 Goinyk Production/Shutterstock.com; p. 7 VectorMine/Shutterstock.com; p. 8 Andrew Mohamed/Shutterstock.com; p. 9 (earthworms) Bukhta Yurii/Shutterstock.com; p. 9 (snail) Marc Andreu/Shutterstock.com; p. 9 (bacteria) paulista/Shutterstock.com; p. 10 Galazveruga/Shutterstock.com; pp. 11, 17 Kichigin/Shutterstock.com; p. 12 chomplearn/Shutterstock.com; p. 13 (fungus cells) Kateryna Kon/Shutterstock.com; p. 13 (plant cells) Tatiana Zorina/Shutterstock.com; p. 14 MiLaN Collection/Shutterstock.com; p. 15 IH82/Shutterstock.com; p. 16 Lidiane Miotto/Shutterstock.com; p. 18 malshkoff/Shutterstock.com; p. 19 (yeast) Cristian Storto/Shutterstock.com; p. 19 (mold) kung_tom/Shutterstock.com; p. 21 Petar B photography/Shutterstock.com.

Cataloging-in-Publication Data

Names: Knopp, Ezra E.
Title: Plant or fungus? / Ezra E. Knopp.
Description: Buffalo, New York : PowerKids Press, 2026. | Series: Fact finders: know the difference! | Includes glossary and index.
Identifiers: ISBN 9781499452785 (pbk.) | ISBN 9781499452792 (library bound) | ISBN 9781499452808 (ebook)
Subjects: LCSH: Plants--Juvenile literature. | Fungi--Juvenile literature.
Classification: LCC QK49.K56 2026 | DDC 580-dc23

Manufactured in the United States of America

CPSIA Compliance Information: Batch #CSPK26. For Further Information contact Rosen Publishing at 1-800-237-9932.

CONTENTS

LET'S INVESTIGATE!

The natural world is filled with living things. Over the centuries, scientists have worked to **classify** the many millions of species on Earth—from plants, to animals, to fungi.

Fungi had been classified as plants up until 1969. This is a good example of the importance of scientific observation and exploration! Even when scientists think they know the facts, they can learn something new that changes their minds. Let's investigate plants and fungi to find out how they are similar and how they are different.

CAN YOU TELL WHICH OF THESE LIVING THINGS ARE PLANTS AND WHICH ARE FUNGI?

FOOD CHAINS

All living things on Earth fit into at least one food chain. This is a connection between living things that shows who eats whom. However, not all living things eat other living things. What roles do you think plants play in a food chain?

EACH ARROW ON A FOOD CHAIN, SUCH AS THIS ONE, SHOWS THE PASSAGE OF ENERGY FROM THE SUN, TO PRODUCERS, AND THEN TO CONSUMERS. THIS FOOD CHAIN SHOWS THAT SOME CONSUMERS ALSO EAT OTHER CONSUMERS.

Plants are producers. They use a **chemical** called chlorophyll to change sunlight and **carbon dioxide** into food for themselves. This process is called photosynthesis. Animals that eat plants, or consumers, receive the energy that plants have stored away as food.

So, what role, if any, do fungi play in the **environment**? How do they fit into a food chain? Do they produce energy, like plants? Or are they consumers, like animals?

Fungi are very similar to consumers in that they "eat" other creatures for energy. However, they are decomposers. They eat, or digest, dead plants and animals and release the energy trapped in them back into the environment. In this way, decomposers are sometimes considered the end of a food chain.

OTHER DECOMPOSERS INCLUDE EARTHWORMS, BUGS, AND BACTERIA. BACTERIA ARE SINGLE-CELLED, **MICROSCOPIC** CREATURES.

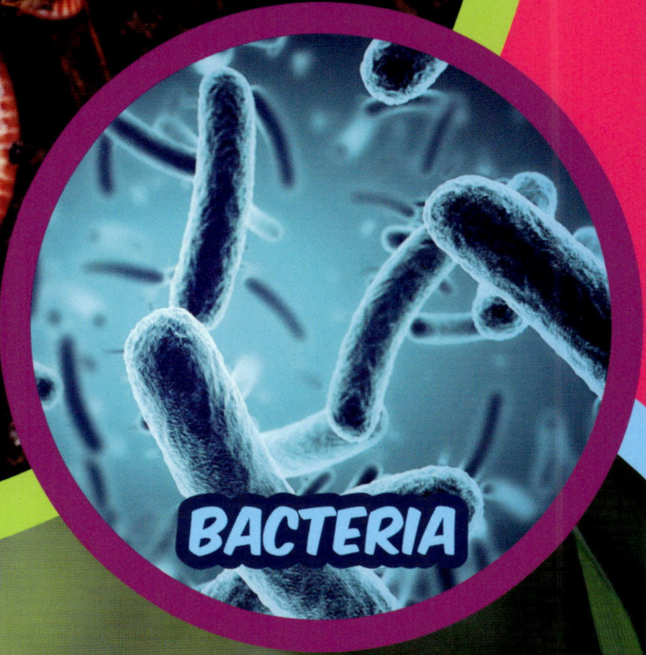

SEEDS AND SPORES

If a **species** can't reproduce, or make more of itself, it will die out. Many plants grow from seeds. Mature plants make tiny grains called pollen, which allow them to make seeds. Birds and bees help spread pollen between flowers.

THIS FUNGUS, COMMONLY CALLED A PUFFBALL, IS RELEASING SPORES FROM A PART CALLED A FRUITING BODY.

Fungi do not make seeds or pollen. They make spores. Spores float on the wind and begin to grow when they land. Spores are a single cell, while seeds are made up of multiple cells.

A CLOSER LOOK

All living things are made up of microscopic parts called cells. Cells carry on life processes. It can sometimes be difficult to tell some plants and fungi apart just by looking at them. Can we tell the two apart by looking at their cells?

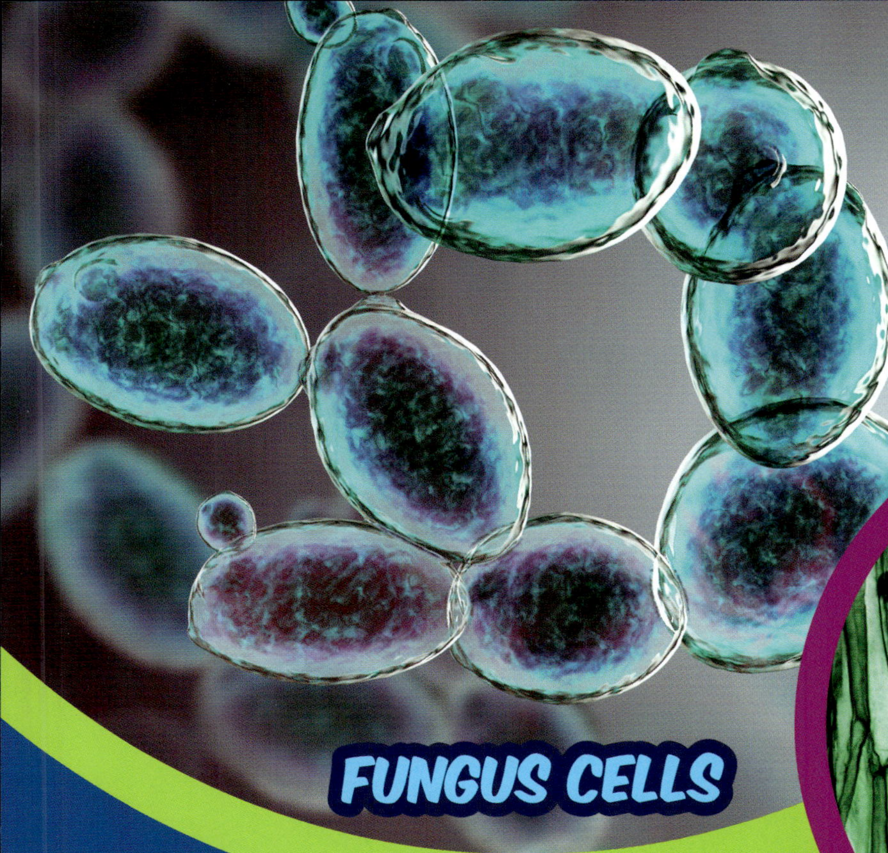

FUNGUS CELLS

PLANT CELLS

CELLULOSE IN THE FRUITS AND VEGETABLES WE EAT IS COMMONLY CALLED FIBER. IT'S VERY GOOD FOR HUMAN HEALTH.

The cells of plants and fungi are very different. Fungi have cell walls made of a hard substance called chitin. Chitin is also what makes up the hard **exoskeletons** of bugs. Plants have cell walls made of cellulose.

FRIENDS?

Fungi and plants can be found growing in the same environments all over the world—from rainforests to deserts. In many cases, fungi have been discovered **infecting** plants! Just as in people, fungi can causes illnesses in plants.

A RELATIONSHIP IN WHICH ONE CREATURE HARMS THE OTHER IS CALLED PARASITIC. A RELATIONSHIP IN WHICH BOTH CREATURES RECEIVE A BENEFIT IS CALLED SYMBIOTIC.

However, not all fungi/plant relationships are bad. In fact many are positive for both. When a fungi and a plant are in the same area, they may share resources. Some fungi connect to plants' roots and give them **nutrients**, and the plants give the fungi sugars in return.

REACHING OUT

Plants have roots that hold them in the ground. Roots take in the water and nutrients plants need to live. Do fungi have roots too?

MYCELIUM IS GOOD FOR THE ENVIRONMENT. IT HELPS DECOMPOSE DEAD MATTER AND RELEASES THE ENERGY STORED INSIDE. HOWEVER, MYCELIUM CAN ALSO INFECT PLANTS AND CAUSE CROPS TO ROT.

Fungi create rootlike threads called mycelium, which makes up the growing body of a fungus. Mycelium branches out as it grows. The threadlike parts are called hyphae. These threads take in water and nutrients from the environment and even create new fungi. While mycelium isn't really the same thing as roots, it serves similar purposes.

WHAT'S IT LOOK LIKE?

Most plants have the same **structure**: roots, stem, leaves, and often flowers. They can be pretty easy to identify. What about fungi? Are they as easy to identify as most plants are?

YEAST IS A SINGLE-CELLED FUNGUS COMMONLY USED IN BAKING. WHEN COMBINED WITH SUGAR, YEAST CREATES THE GAS CARBON DIOXIDE. THIS REACTION IS USED TO MAKE BREAD AND OTHER FOODS.

Fungi can take on a number of different-looking structures. Most mushrooms look similar: a stem and a cap. But did you know that mold is a fungus? Mold spores land on food and get nutrients from it. As the mold grows, the fruit rots.

KEEP INVESTIGATING!

Now you know several scientific ways to tell plants and fungi apart. Maybe you've used your eyes to compare their structures. Maybe you've also used scientific tools, such as a microscope, to make comparisons between the two. Either way, you've used your powers of observation and investigation to discover the differences between plants and fungi.

But why stop here? There's so much more you can learn about plants, fungi, and the natural world. Keep investigating to learn even more!

DID YOU KNOW THERE ARE PLANTS THAT EAT MEAT? DID YOU KNOW SOME FUNGI GLOW IN THE DARK? IT'S TRUE!

GLOSSARY

carbon dioxide: A gas that animals breathe out and which plants need to make food.

chemical: Matter that can be mixed with other matter to cause changes.

classify: To sort objects into different classes.

environment: The conditions that surround a living thing and affect the way it lives.

exoskeleton: The hard outer covering of an animal's body.

infect: To make someone sick.

microscopic: So small it can only be seen with a tool called a microscope, which uses lenses to make tiny things appear bigger.

nutrient: Something taken in by a plant or animal that helps it grow and stay healthy.

species: A group of plants or animals that are all the same kind.

structure: The arrangement of parts of a living thing.

FOR MORE INFORMATION

BOOKS

Bonkoski, Ariel. *Mushrooms & Fungi for Kids: An Introduction to Mycology*. Cambridge, MN: Adventure Publications, 2025.

DK. *Plants and Fungi!* New York, NY: DK Children, 2023.

Rossiter, Brienna. *Fungi*. Mendota Heights, MN: North Star Editions, 2025.

WEBSITES

Fungi
science4fun.info/fungi
This website offers even more information about fungi.

Plants
science4fun.info/plants
Science4Fun features a collection of facts about all kinds of plants too!

INDEX